Science Investigations

ELECTRICITY:
AN INVESTIGATION

JOHN FARNDON

PowerKiDS press.

New York

Published in 2008 by The Rosen Publishing Group, Inc.
29 East 21st Street, New York, NY 10010

First Edition

The publishers would like to thank the following for permission to reproduce these photographs:
Corbis: 5 (DK Limited), 10 top (Vauthey Pierre/Sygma), Cover and 16 (Ethan Miller/Reuters), 24 right (© H. David Seawell), 28 (Michael S. Yamashita); Last Resort Picture Library: 13, 17, 21, 29; OSF/Photolibrary: 6 (Warren Faidley), 12 (Workbook, Inc.), 18 right (Mary Plage), 20 (Index Stock Imagery), 23 (Index Stock Imagery); REX Features: 18 left; Science Photo Library: 4 (Charles D. Winters), 7, 15 (Peter Menzel), 26 left (Gusto), 26 right (Martyn F. Chillmaid); Topfoto: 10 bottom (The Image Works), 14, 22, 24 left.

Editors: Sarah Doughty and Rachel Minay
Series design: Derek Lee
Book design: Malcolm Walker
Illustrator: Peter Bull
Text consultant: Dr. Mike Goldsmith

Library of Congress Cataloging-in-Publication Data

Farndon, John.
  Electricity : an investigation / John Farndon. — 1st ed.
      p. cm. — (Science investigations)
  Includes bibliographical references and index.
  ISBN-13: 978-1-4042-4287-6 (library binding)
  1. Electricity—Juvenile literature. I. Title.
  QC527.2.F374 2008
  537—dc22
                                        2007032605

Manufactured in China

Contents

What is electricity? 4

How does electricity make sparks? 6

How are electrons and protons charged? 8

What are the effects of static electricity? 10

How does electricity make things glow? 12

How do we measure electrical charge? 14

How does electrical charge discharge? 16

How does a battery work? 18

How do we make a battery with a bigger charge? 20

How do we make an electrical current? 22

How does electricity flow in different materials? 24

How do we connect circuits? 26

How are electricity and magnetism linked? 28

Glossary 30

Further information 31

Index and Web Sites 32

What is electricity?

Electricity is the amazing energy that we use for countless different tasks. We use it to make electric lights shine and mircowave ovens cook meals. We use it to make music on a CD player or create a TV picture. We use it to power everything from the minute workings of a computer, to the mighty engines of a high-speed train.

Almost all of the electricity we use is artificially produced by generators and batteries. Yet electricity is one of the basic natural forces of the universe. Did you know there is natural electricity all around us, all the time? Did you know, in fact, that there is electricity in every single atom in every substance? How do we know it is there, and how can we take advantage of this universal energy? One sign of naturally occurring electricity, called static electricity, is in the way it can make things stick together.

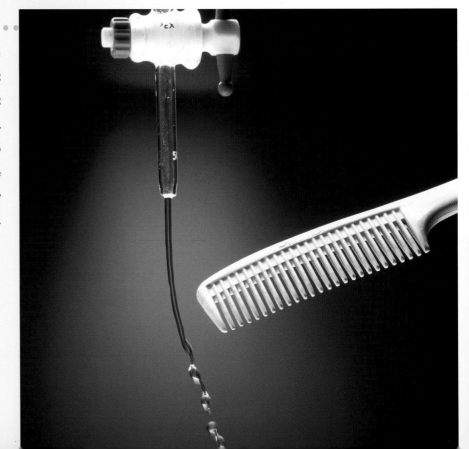

You can bend water with electricity. Turn on a faucet with the smallest, steadiest flow you can get, then run a comb through dry hair or rub it with a cloth. Now hold the comb close to the water flow without touching it.

INVESTIGATION

How does electricity make substances stick together?

MATERIALS

A woolen sweater and a party balloon.

INSTRUCTIONS

Rub a balloon on your hair or sweater. You will find that you can now stick it to a wall. Time how long it stays before it falls down.

Repeat this in the bathroom, just after someone has taken a hot, steamy shower. (In the bathroom, water in the air and on the walls helps electricity move away from the balloon more quickly. Similarly, electrical charges are stronger in the dry air of a centrally heated house, than in a more damp house.)

FURTHER INVESTIGATION

What other substances can you make stick together after rubbing in the same way?

Do they all behave the same way in every season of the year?

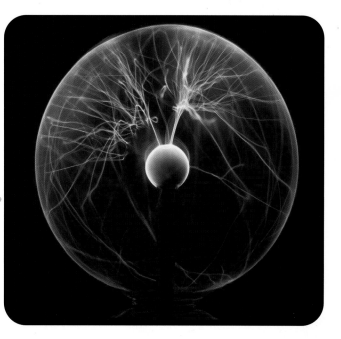

A plasma ball is a special glass ball in which you can actually see glowing streams of electricity. You can see electrical attraction in action as electricity is drawn instantly to wherever you touch the ball's surface.

How does electricity make sparks?

There is no more dramatic evidence of natural static electricity than lightning. Lightning is the brilliant flash of light that occurs during a thunderstorm. It is actually a gigantic surge of electricity leaping between a thundercloud and the Earth. Inside thunderclouds, violent air currents hurl tiny water droplets and ice crystals together with such force that they become electrically charged.

The electrical charge builds up and up inside the cloud until it overflows and "discharges" in a flash of lightning. Sometimes, lightning flashes within the cloud, making the whole cloud glow. This is called *sheet lightning*. Sometimes, however, it makes brilliant, crooked rivers between the cloud and the ground. This is called *forked lightning*. Why do you think the cloud's electrical charge ends up going where it does?

Lightning occurs when an electrical charge builds up inside a thundercloud as water drops and ice particles are hurled together by violent air currents. What do you think this has in common with the balloon sticking to your sweater on page 5?

One of the safest places to be if you're out in the countryside in a thunderstorm is inside a car. A metal box such as a car channels the electricity around the outside, leaving you safe inside. Here, a scientist is demonstrating this with a special metal cage called a Faraday cage.

INVESTIGATION

How can you create a lightning spark?

MATERIALS

A large iron or steel pot (not aluminum) with a plastic handle, rubber gloves, an iron or steel fork, and a plastic sheet.

INSTRUCTIONS

Tape a plastic sheet to a tabletop. Put on the rubber gloves. Hold the large iron pot or pan by its insulating handle, and rub the pan vigorously to and fro on the plastic sheet.

Holding the fork firmly in the other hand, bring its prongs slowly near the rim. When the gap between pot and fork is small, a tiny spark should jump across (if you darken the room you may see the spark more clearly).

FURTHER INVESTIGATION

If your house has carpets made from artificial wool, try taking your shoes off and rubbing your bare feet several times across the carpet. Now, almost—but not quite—touch a metal door handle, or a metal. What happens? Beware, you may get quite a shock, but it won't do you any harm.

How are electrons and protons charged?

An atom is made up of protons, neutrons, and electrons. The nucleus of an atom is made up of protons, which have a positive charge, and neutrons, which are neutral. The electrons, which have a negative charge, orbit the nucleus.

Every substance in the Universe is made of minute parts called *atoms*, and electricity is created by even tinier bits of atoms called *electrons*. In the middle or nucleus of every atom there are other parts called *protons*. Both electrons and protons are electrically charged, which means they have a special kind of energy.

Electrons are said to be have a "negative" charge, which means their charge pulls them toward protons, which have an opposite, "positive" charge. An electron's negative charge also means it is pushed away by other electrons, which also have a negative charge. Usually, electrons are held closely to the atom by their attraction to protons. Electrical effects, such as electrical attraction, occur when electrons at the edge of the atom break loose or are transferred to other atoms.

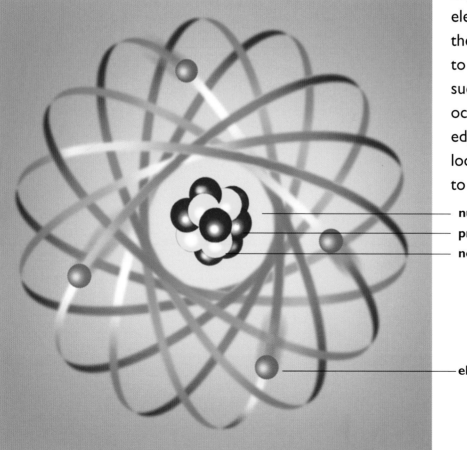

nucleus
proton
neutron

electron

INVESTIGATION

Investigate electrical attraction and repulsion

MATERIALS

A roll of plastic tape, a plastic mug, and scissors.

INSTRUCTIONS

Stick the end of a piece of tape to a plastic mug. Cut off another piece and move toward the first piece. Both pieces curl away—they repel each other. Lightly stroke one piece of tape between your finger and thumb several times. Now try moving the two pieces of tape together. This time they jump toward each other. They are attracted.

As each piece of tape is pulled off of the roll, it pulls some electrons with it. The electrons give the tape a negative charge of electricity. Both pieces of tape have the same negative charge so they repel each other. Stroking the tape neutralizes its charge so that the unstroked, charged piece of tape is drawn toward it. A charged object is always drawn toward an uncharged object, because there is a difference in charge.

FURTHER INVESTIGATION

When you brush a comb through dry hair, the hair often lifts up to stick to the comb. What could be attracting the hair to the comb?

Do different kinds of comb or brush make any difference?

What are the effects of static electricity?

Gilders create beautiful designs in gold by skillfully applying very thin sheets of gold called *gold leaf* with a brush. Why do you think they rub the brush through their hair before they pick up each piece of gold leaf?

As we have seen on earlier pages, static electrical effects are created by the movement of minute particles called electrons. Electrons are negatively charged. So, as electrons pile up in one place or move away from another, an imbalance of electrical charge is created. The range of electrical effects created this way is huge. Lightning is the most dramatic example, but there are many others, some wanted, some unwanted.

Among the unwanted effects are potential damage to sensitive electronic equipment, such as computer chips. The static charge built up on people's bodies as they simply move around can often be enough to burn a hole in a microchip. This is why engineers working with electronic parts are very careful to reduce static: they use antistatic sprays, keeping the air moist, and wear "antistatic bracelets" that help drain the static charges. How do these methods deal with unwanted static?

Electricity and fuel are a dangerous combination. A small spark of electricity can easily make fuel explode when a tanker is unloading. For this reason, all fuel tankers have devices for draining any static electricity.

10

INVESTIGATION

How do you make a simple photocopier?

Photocopiers copy pictures by shining a reflection of the picture onto a special roller. This creates a matching pattern of electrical charges on the roller's sensitive surface. The charges make ink dust stick to the roller in exactly the same pattern. So when the inked roller is rolled across paper, it prints out a perfect copy of the picture.

MATERIALS

A sheet of dark plastic (such as a plastic document wallet), talcum powder, tape, and a cotton cloth.

INSTRUCTIONS

Lay the plastic sheet on a table and stick some tape on it in a pattern in the shape of your initials. Sprinkle talcum powder on the cloth and rub it in. Peel the tape off of the plastic, and shake the cloth just above the plastic to shake out the talcum powder. The talcum powder sticks only to the places where you have peeled tape off. Why do you think this might be? (Clue: it is not because the tape leaves glue behind.)

FURTHER INVESTIGATION

Your clothes sometimes feel hard when they come out of the washing machine. This is because the fibers in certain fabrics, especially synthetics, rub together in the machine and exchange electrons. An electrical charge builds up on the fabric, making the fibers cling together so that they feel slightly stiff.

When you add fabric conditioners, they lubricate the fibers and reduce the buildup of electrical charge. The fibers slide more easily over each other, and the fabric feels softer. Hold a magnifying glass over the fibers of clothes washed using fabric conditioner and clothes washed without it. Can you see any difference?

How does electricity make things glow?

The brilliant glow of a lightning flash shows just how brightly a charge of electricity can make air glow. If you look around you, you can see how it can make solids glow, too. In fact, a city at night glitters with all kinds of glow made by electricity—neon lights on displays, street lamps, fluorescent lights in stores and offices, car headlights, and much more.

If you go near the North Pole, you may see the night sky glowing with the spectacular northern lights or aurora borealis. The air glows in different colors where atoms are bombarded by electrically charged particles that stream from the Sun. Why do you think different parts of the air glow different colors?

All these electric glows occur because electricity energizes atoms or gets them, as scientists say, "excited," so that they radiate little bursts of light. The atoms are excited by being bombarded by electrically charged particles—electrons. When you see a light glowing—except where there is a flame—you know the glow comes from atoms excited by an electrical charge. See if you can guess where the electrical charge comes from, and what atoms are glowing in each kind of light you see.

INVESTIGATION

How do you make a light bulb glow with static electricity?

MATERIALS

A hard rubber comb and a fluorescent light bulb (not one with a glowing wire"'filament" inside).

INSTRUCTIONS

Go into a dark room. Charge the comb on your hair or a sweater. Make sure to build up a lot of charge. Touch the charged part of the comb to the light bulb and watch carefully. You should be able to see very small sparks. Try touching different parts of the bulb. What causes the sparks?

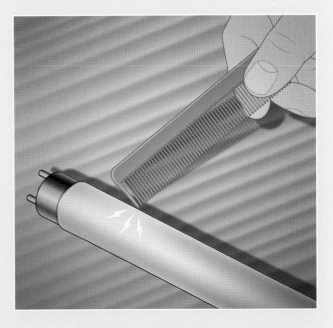

FURTHER INVESTIGATION

The neon lights you can see flashing in downtown areas are glass tubes filled with neon and other gases, and the glow is created as atoms of neon are excited by an electrical charge. Neon glows pure red, so how do they get other colors?

Every time you turn on a TV or a computer, you are making an electric glow. The screen glows as it is bombarded from inside by electrons fired from a special gun, and the picture is built up as the gun scans rapidly to and fro.

How do we measure electrical charge?

Sometimes it is very easy to detect a static electrical charge. No one can miss a flash of lightning, for instance. Even small charges are easy to spot when they create a spark, or a glow, as we saw on page 13. But it is not always quite so easy to tell when an electrical charge has built up. Often, you only discover an electrical charge when it discharges—for instance, if it makes your hand tingle when you touch a metal handrail. Until then, you don't know the charge is there.

To detect invisible charges, and to measure just how big they are, what you really need is a device called an *electroscope*. This works because electric attraction can make things move or bend. So all electroscopes include an indicator that moves or bends when an electric charge is detected. The amount the indicator moves or bends shows how strong the charge is.

. . . .

When trees are hit by lightning, they can split apart because their sap boils instantly. The heat needed to create this effect is dramatic evidence of the huge amount of electricity involved in lightning—many millions of volts.

INVESTIGATION

How do you make a simple electroscope?

MATERIALS

A 12 in. (30 cm) square of aluminum kitchen foil, tape, a wide-mouthed jar, a strip of "silver" foil from a candy bar, a large needle, thin cardboard, a comb, and 6 in. (15 cm) of bare, stiff wire.

INSTRUCTIONS

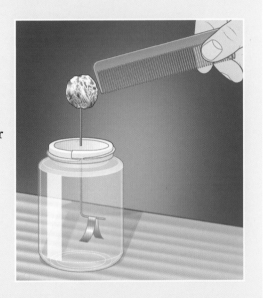

Cut a circle from cardboard slightly wider than the jar's mouth. Punch a hole in the middle of the circle with a needle (ask an adult to help). Pass the wire into the hole. Crumple the kitchen foil into a tight ball, and push it on one end of the wire. Make a hook on the other end of the wire. Take the silver foil, fold it in half, and hang it over the wire hook. If the ends do not hang side by side, press them together gently. Carefully lower the circle with its wire and foil into the jar. Tape the circle in place to cover the jar. Bring a charged comb near the foil ball on top of the jar. The two halves of the silver foil should curve away from each other.

Van der Graaf generators build up static electrical charges by rubbing a fabric belt over metal rollers. The charge from the generator makes this boy's hair stand on end as the charge in each hair repels the charges in the other hairs. What does the straightness of his hair tell you about the size of the charge?

FURTHER INVESTIGATION

Use your electroscope to detect the strength of charges on different materials that you rub together. Does wool hold a stronger charge than nylon?

How does electrical charge discharge?

When you rub a balloon on a sweater, electrons rub off the surface of the balloon and onto the sweater. So the sweater becomes electrically charged. "Charged" is an old-fashioned word for "filled," so saying it is charged is just another way of saying extra electrons have piled up. Just as a glass will overflow when it is filled with water, so electrical charge is likely to spill over or "discharge" when too many negatively charged electrons build up in one place. But although water spills freely in all directions, electrical charge flows only to places where there is a lack of negative electrons, and tends to follow narrow paths. You can see this very clearly in the jagged paths taken by forked lightning through the sky. Here we look at how electrical discharges find their path.

Forked lightning occurs when a build up of charge in a thundercloud discharges to the ground, which can soak up the excess electrons. Lightning strikes not just anywhere on the ground—it is focused on tall buildings and isolated trees. Why do you think this might be?

Look at the spark of the igniter on a gas stove and you will see it is not like flame. Instead, it makes a short, jagged line between two points. What do you think these two points have in common with a thundercloud and a tall building?

INVESTIGATION

See the lines of charge around a plasma ball

MATERIALS

An inexpensive plasma ball (available from a toy store or an electronics shop) and a small, low-voltage, fluorescent light tube.

INSTRUCTIONS

Turn on the plasma ball and allow the charge time to build up. Place your fingers on the ball and see how the lines of charge flow toward your fingers. Move your fingers around and see how the lines of charge follow your fingers. Why do you think this happens?

FURTHER INVESTIGATION

Place one end of the fluorescent light bulb against the plasma ball so that it lights up. Hold the bulb at the end farthest from the ball and the entire bulb should light up. Now if you hold the bulb in the middle, only half of it will light up. Why do you think this might be?

How does a battery work?

Most of the electrical charges described so far are created by two things rubbing together. But electrical charges can be created by chemical reactions, too. This is how batteries work. Batteries come in all shapes and sizes. Some are as small as a pill or as thin as a sheet of paper. Others are as large as a refrigerator. All batteries create an electrical charge in the same way: two chemical substances react together, and electrons flow from one substance to the other. In an ordinary flashlight battery, for instance, these substances are zinc and carbon, but in a car battery, they are lead and lead dioxide.

If you have fillings and you accidentally bite the foil wrapper when you eat a piece of chocolate, you sometimes get a nasty taste and a slight tingling feeling in your teeth. Why do you think this is?

Scientists often learn how things work by looking at other things. What could you learn about electric batteries by looking at a dam holding up water?

INVESTIGATION

How do you make a battery with lemons?

MATERIALS

A flashlight bulb and bulb holder (from an electrical shop), about 12 in. (30 cm) of steel wire and another 12 in. (30 cm) of copper wire, wire cutters, lemons, and a knife.

INSTRUCTIONS

Make the lemons as juicy as possible by rolling them firmly on the table. Then cut one lemon in half. Strip any plastic coating off, about $1/10$ in. (25 mm) from the ends of the wires with the wire cutters (ask an adult to help). Push one end of each wire into opposite sides of one of the lemon halves. Now attach the other ends of the wires to the bulb holder. Does this make the bulb light up?

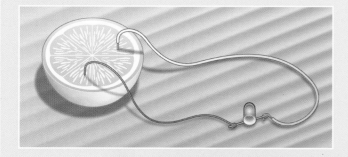

If the bulb doesn't light, try connecting more lemons in the same way, making sure each lemon half has one copper and one steel wire.

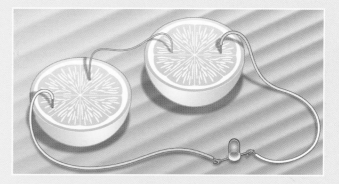

The acidity of the lemon juice is making the metal in the two wires react. The chemical reaction strips the positive *ions* (atoms with electrons missing) from the steel, and it leaves it with an excess of negatively charged electrons. This sends an electrical *current* (a flow of electrons) from the steel wire to the copper wire. The copper wire draws positive ions from the acid lemon juice and becomes positively charged. The bulb lights up as the stream of electrons flows through it.

FURTHER INVESTIGATION

Try the same experiment with a jar of vinegar, instead of lemons. What do lemon juice and vinegar have in common? Are there any other substances you could use?

How do we make a battery with a bigger charge?

Not all batteries are useless when they lose their charge. Many batteries, such as those in MP3 music players, are rechargeable. To recharge, you connect the battery to an electricity supply for awhile. This draws electrons back to the positive terminal and charges the battery again.

In any battery, the part that creates the charge is called the *cell*. Most batteries, such as those in flashlights, have just a single cell. Some, such as those in cars, are made from many cells. The more cells a battery has, the bigger the charge it can create. Making a circuit by connecting a loop of wire to both sides, or *terminals*, of a battery cell allows you use its electricity. To make a bigger charge, you simply connect other cells into the loop, like extra links in a chain.

The electricity in any cell is concentrated at two points or surfaces called *electrodes*. One, called the *negative electrode*, has more electrons; the other, called the *positive electrode*, has fewer. The electricity is created by the difference in charge between them. Both electrodes are dipped in a special paste or liquid, called the *electrolyte*. Whenever the battery is connected into a circuit and the current starts to flow, electrons migrate from the positive to the negative through the electrolyte. Eventually, so many electrons migrate that the charge difference between the electrodes is gone and the battery goes flat.

INVESTIGATION

How do you build a multicelled battery from coins?

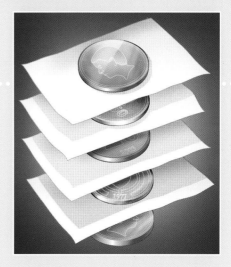

MATERIALS

Silver and copper coins (it is important that the metals are different), strips of paper towel, lemon juice, and a bowl.

INSTRUCTIONS

Soak the paper towel strips in lemon juice to form the barrier between the coins. Sandwich a strip of lemon-soaked towel between a silver and a copper coin. Add more and more sandwiches of silver coins, lemon strips, and copper coins on top, turning over the sandwich each time, so that the same two coins never touch. Wet your finger and thumb, and carefully try to pick up the pile between them. How can you tell the money pile is electric? (Clue: you should feel a slight tingling.)

The lemon juice is the electrolyte, the coins are the two electrodes. The coins and lemon combine to make a multicell battery.

FURTHER INVESTIGATION

Try making a multicell battery with other metal objects or other electrolytes. The best electrolytes are acidic or salty solutions—for example, salt dissolved in cold water. The crucial thing about the metals is that they need to be different. You could use copper wire and galvanized screws (zinc). You can make a single-cell battery with a jar of salt water with your copper and zinc objects dipped in for each of the terminals. You can make a multicell battery with a series of jars, with the copper terminal in each connected by wires to the zinc terminal in the next.

In a car battery, each of the cells is filled with sulfuric acid, which forms the battery's electrolyte. Because the acid is liquid, it is called a *wet cell battery*. How many cells do you think this battery has?

How do we make an electrical current?

Static means "not moving" and *static electricity* gets its name because the charge builds up in one place. Sometimes a static charge does move, but usually in a single jump. The movement of charge can then be huge, as in lightning, but it is so brief and uncontrollable that it is hard to use as energy. But as we have seen, an electrical charge follows a particular path when it discharges, and this path is always the one that is easiest for the electricity to move along.

Electrical charges move more easily through some substances than others. By providing a line of material through which electrical charge can move easily, such as a wet string or a piece of copper wire, you can guide an electrical charge to particular places. If you make a complete loop with this line of material, you can set up a continuous flow of electrical charge, called a *current*. But just as a line of people waiting to get into a movie theater cannot move until the doors are open, so an electric flow cannot begin until the loop is complete, and the path for the charge opens up.

In this game, you have to have a steady hand as you conduct "surgery," and extract organs carefully without touching the body. If your "forceps" or the organ touches the body, for an instant a light or buzzer goes on and you have failed. Why do you think the buzzer or light goes on when you touch the body with the forceps or the organ? (Clue: the body is connected to a battery.)

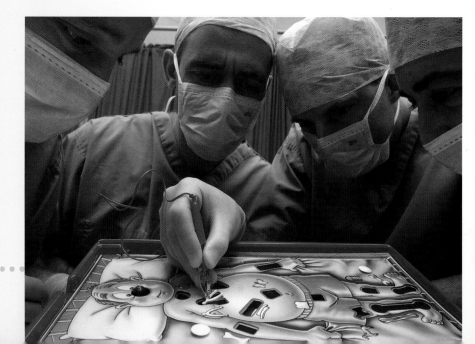

22

This fountain is kept flowing by a pump that keeps the water going around and around—drawing in the water that falls into the basin and pushing it back up. If the pump is turned off or its intake becomes blocked, the fountain stops. What does this tell you about an electric current?

INVESTIGATION

How do you make an electric circuit?

MATERIALS

A 1.5 V battery with metal flanges for attaching clips to, as illustrated, two pieces of wire with clips at both ends (from an electrical store), a small 1.5 V bulb, and a bulb holder.

INSTRUCTIONS

Clip one end of one wire to one flange or terminal of the battery. Clip the other end to one terminal of the bulb holder. Clip one end of the other wire to the other terminal of the battery. Clip the other end to the other terminal of the battery. Why should the bulb now light up? How would you stop the bulb from lighting up? Why would this work?

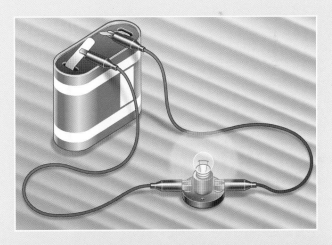

FURTHER INVESTIGATION

Try taking the battery out of the circuit and connecting the wires and bulb together in a loop. The bulb, of course, is now unlit. But what does this tell you about electric circuits?

How does electricity flow in different materials?

Some substances transmit, or "conduct," electricity much more easily than others. The best conductors are metals such as copper. This is because metal atoms always have a few electrons loosely held in place, so there are lots of "free" electrons to transmit the current. Water also conducts electricity well, for the same reason—which is why it is very dangerous to go near electrical equipment with wet hands. If your hands are wet, the water can seep into the switch and make instant contact with the electrical circuit inside, giving you a serious or even fatal shock. Other materials do not conduct electricity because their electrons are not free to move. These materials are called *insulators*. Air, plastic, and rubber are good insulators.

Electrical plugs vary around the world: some have two wires, one for each half of the circuit; some have an extra "earth" wire to allow any excess current to flow away safely. If you ask an adult to undo a plug, you'll see the wires inside are made of copper, covered in plastic. Why do you think wires are made of copper but coated in plastic?

Here an electrician is working on an overhead cable that carries a huge amount of electric power. The glass and ceramic doughnuts you can see are insulators to stop the electricity from escaping into the poles that hold the cable up.

What substances are more electrically resistant?

When electrons bump into atoms in a conductor, the current is reduced. This is called *resistance*.

MATERIALS

A 1.5 V battery with metal flanges for attaching clips to, three pieces of wire with clips at both ends (from an electrical store), a small 1.5 V flashlight bulb and bulb holder, and materials to test for resistance (e.g. scissors, wooden, metal, and plastic rulers, pencil, eraser).

INSTRUCTIONS

Connect one end of a wire to the battery and the other end to one of the bulb flanges. Now connect one end of a second wire to the other bulb flange. Connect one end of the third wire to the other battery terminal. You are left with two free clip ends. See how brightly the bulb shines if you simply connect these ends. Now connect the free clips to each side of the material to be tested. The dimmer the bulb glows, the more resistant the material is.

Which materials are the most resistant? Does the shape of the material make a difference?

FURTHER INVESTIGATION

Soak a wooden pencil in water overnight. Ask an adult to use a scalpel or craft knife to split the pencil in half lengthwise, leaving the graphite (the "lead") in one half of the pencil. Set up the circuit as in the investigation, then connect one free clip to the point of the graphite. Touch the second free clip on the graphite some of the way down to complete the circuit and make the bulb light up. Move the second clip backward and forward along the graphite, and watch the bulb growing brighter or dimmer: the shorter the circuit, the brighter the light will be.

How do we connect circuits?

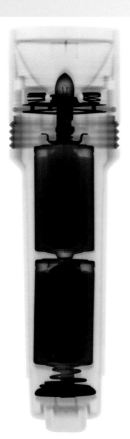

There are many different kinds of electrical circuits. Some are minute, such as the tiny circuits etched on the circuit boards inside a computer. Others are gigantic, such as the massive cables that carry electric power from power stations to your home. Large or small, all circuits have the same three elements. First, there is a *conductor* through which the electricity flows. Second, there is a *load*, which is the equipment the electricity is powering, such as a TV. Third, there is the *energy source*. This can be either a battery or a generator, but each works by building up more electrons in one part of the circuit.

Like every piece of simple electric equipment, a flashlight is made by combining three elements—a load, a conductor, and an energy source. Looking at this X-ray, which do you think is which?

Look at different batteries, and you will see that they are often marked with a figure such as 1.5 V or 1.5 volts. Usually, the bigger the battery, the bigger the number of volts and the more energy it can provide. Volts are said to be a measure of "potential difference"—but what difference does the battery create?

How do you make series and parallel circuits?

MATERIALS

A 1.5 V battery, connecting wires and clips, and three bulbs.

INSTRUCTIONS

To make a series circuit, connect the battery and wires in a single loop with one bulb. Add the other two bulbs into the loop. The bulbs should glow equally brightly.

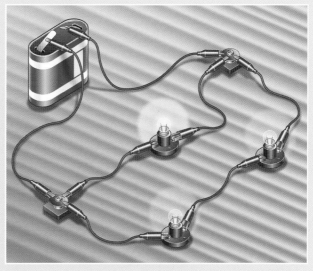

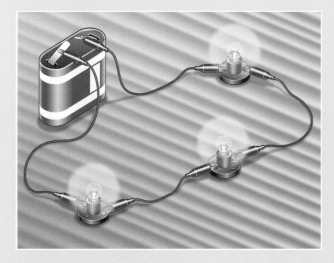

To make a parallel circuit, use the connectors to split the circuit into two branches, with one bulb in each branch. Each branch takes half the current, so the bulbs should glow equally brightly. Add a second bulb to one branch. They should glow dimmer than the other, because the two bulbs create more resistance on this branch.

Electrical circuits must form a complete loop, but they can be connected in various ways. If all the components are in a single loop, the circuit is said to be in *series*. If the circuit splits into branches, it is said to be connected in *parallel*.

FURTHER INVESTIGATION

Try connecting different-sized batteries into your bulb circuit. Try varying the size or number of batteries and the way in which they are connected.

How are electricity and magnetism linked?

If you put a small magnetic compass near a wire carrying an electric current, a surprising thing happens. Instead of pointing north, the compass needle swings around to point toward the wire, just as if it was next to a strong magnet. In fact, that is just what it is. Every electric current creates its own magnetic field, not so different from that around a steel magnet. Indeed, electricity and magnetism are opposite sides of the same coin, and they always occur together. Wherever electricity flows, there is magnetism. Electricity can be used to make magnets and magnets can be used to generate an electric current. It is these two effects—electricity creating magnetism and magnetism creating electricity—that we investigate here.

Electricity can be used to create strong magnets called *electromagnets* whose magnetism can be turned on and off with the electric current. This powerful magnet is used to lift scrap metal in scrapyards. You can see how the magnetism stops and the metal drops when the current is turned off.

INVESTIGATION

Is electricity magnetic?

MATERIALS

A ball of fine steel wool, an old pair of scissors, a 1.5 V battery, and connecting wires and clips.

INSTRUCTIONS

Lay down the batteries and the wire ready to make a circuit, but don't actually connect the wires to the battery. Snip tiny shreds of steel wool off of the ball and scatter them over the circuit. Connect the wires to the battery terminal and watch what happens to the

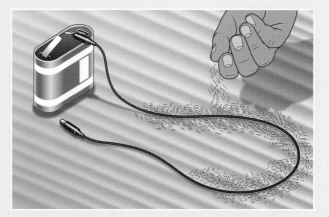

clippings. What happens when you disconnect the battery?

FURTHER INVESTIGATION

What happens to the magnetic effect if you arrange the wires into different patterns? Try winding them into a coil, for instance.

You don't need a battery to provide the electricity for a flashlight bulb. To make this flashlight work, you simply pump the handle. The pumping action moves a magnet rapidly around and around inside a coil of wire to generate electricity.

Glossary

Antistatic device
A device designed to drain off unwanted static electricity.

Atom
The smallest part any pure substance can be broken down into. Atoms are much too small to see except with extremely powerful microscopes.

Battery
A package that provides a source of electrical energy. Batteries consist of one or more cells.

Cell
The basic unit of a battery, consisting of a positive electrode and a negative electrode.

Charge
The electrical attraction of an object or atomic particle.

Circuit
An unbroken loop in which an electric current flows.

Conductor
Any material that allows electricity to flow easily through it.

Current
The flow of electrical charge that is created when electrons move.

Discharge
When a buildup of electrical charge is suddenly released, as when the electrical charge in a thundercloud flashes to the ground as lightning.

Electrode
The point in a battery where charge is concentrated.

Electrolyte
The conducting substance in which the two electrodes of a battery are dipped.

Electromagnetism
The combined effects of electricity and magnetism.

Electron
A tiny particle, much smaller than an atom. It is negatively charged and spins at high speeds around the nucleus of an atom.

Electroscope
A device for detecting or measuring static electricity.

Generator
A machine for making electricity by moving magnets.

Insulator
A material that restricts the movement of electricity.

Ion
An atom or group of atoms with a positive or negative electrical charge.

Negative charge
The natural attraction an electron has to an atom, or anywhere lacking in electrons. A negative charge can also be just a surplus of electrons.

Nucleus
The central part of an atom, containing positively charged protons and neutral neutrons.

Positive charge
A lack of electrons. Any surface or atom that lacks electrons attracts negatively charged electrons to it.

Resistance
The ability of a substance to resist the flow of electricity through it.

Static electricity
Electricity that stays in one place: the electrical charge involved in lightning and on the surface of rubbed objects.

Volt
The unit for measuring how much electrical force there is driving an electric current. It depends on the difference in the number of electrons in two parts of an electrical circuit.

Further information

BOOKS

Batteries, Bulbs, and Wires (Young Discoverers: Science Facts and Experiments)
by David Glover
(Kingfisher, 2002)

Electricity (DK Eyewitness Books)
by Steve Parker
(DK Children, 2005)

Electricity (Science Alive!)
by Darlene Lauw and Lim Cheng Puay
(Crabtree Publishing Company, 1997)

Electricity (Science Experiments)
by John Farndon
(Benchmark Books, 2000)

Electricity and Magnetism (Prentice Hall Science Explorer)
by Prentice Hall
(Pearson Prentice Hall, 2006)

Electricity and Magnetism (Usborne Understanding Science)
by Peter Adamczyk and Paul-Francis Law
(EDC Publishing, 1994)

Electric Mischief: Battery-Powered Gadgets Kids Can Build (Kids Can Do It)
by Alan and Lynn Bartholomew
(Kids Can Press, 2004)

CD-ROMS

Eyewitness Encyclopedia of Science
Global Software Publishing

I Love Science!
Global Software Publishing

ANSWERS

page 5 Any substance with artificial fibers such as nylon and lycra, and also various hard plastics and polythenes (such as garbage bags) may stick together with static electricity—especially in the summer, when the air is drier.

page 6 Rubbing and knocking can create static electricity. You create electricity by rubbing the balloon on your sweater. The droplets in a thundercloud knock together as they swirl around.

Opposite electrical charges are attracted to each other. So the charge that builds up in the cloud is drawn to the nearest place with an opposite charge. This is often a high point on the ground.

page 7 The door handle or metal will tingle with static electricity.

page 9 When you brush a comb through your hair, you knock electrons off the hair onto the comb. The hair is then attracted to the electrically charged comb.

The effect is much more obvious with plastic combs and brushes than with metal or natural bristle combs.

page 10 Gilders rub the brush through their hair to charge it with electricity so that the gold leaf sticks to it.

Static builds up when electric charge cannot flow away. Moisture in the air and antistatic sprays are conductive and help the charge flow away. Antistatic bracelets help the charge flow to the ground.

page 11 Peeling the tape off of the plastic leaves an area charged with static electricity. This attracts the powder, creating an image.

The fibers on the clothes washed without fabric conditioner look more matted because static electricity makes them stick together.

page 12 Every substance glows a certain color. Different colors are created when the particles hit different gases in the air.

page 13 When the charged comb touches the bulb, electrons move from it to the bulb, causing the small sparks. In a similar way, electrons to light a light bulb come from the electrical power lines through a wire in the end of the tube.

With other gases (for example, argon gives lavender, krypton gray-green, xenon blue-gray, and mercury vapor light blue).

page 15 The straightness of the hair means each hair is being pushed as far apart as possible, so the charge is strong.

Nylon and many artificial fibers hold a stronger charge than wool or other natural fibers.

page 16 Lightning is focused on tall buildings and isolated trees because it is attracted to the nearest point to the cloud.

page 17 Both are the nearest point where the static electricity can discharge.

The lines of negative charge are drawn to your uncharged fingers.

The bulb lights up as charge flows through it to discharge into your hand. The charge only flows and the bulb glows only as far as your hand.

page 18 When you bite the foil, it creates a chemical reaction between your fillings and the metal, which creates a nasty taste and the small electrical charge that tingles.

The more water a dam holds, the stronger the current it creates when it is let out. The bigger the charge in a battery, the stronger the current it can give.

page 19 Lemon and vinegar are both sour-tasting acids. Their acidity is what helps create the chemical reactions that make electricity.

page 21 Six. The yellow filler caps allow the electrolyte to be topped up in each cell.

page 22 The body is connected to the same electric battery as your forceps, so every time you touch the body with the forceps, you are completing an electric circuit.

page 23 A circulating current needs driving force to drive it round. The water feature needs the pump. An electric current needs a battery.

The bulb lights up because the circuit is now completed.

Take the clip away from the battery terminal. This breaks the circuit and stops the electric current flowing.

There must be something to push the electrons that carry the charge through the circuit. This is called the *electromotive force* and is typically provided by a battery or a generator.

page 24 Wires are made of copper because copper conducts electricity very well. They are coated in plastic because plastic is a good insulator and stops electricity being lost.

page 25 Some materials, such as wood and rubber, offer a great deal of resistance. This diminishes the current and makes the light glow dimly or not at all. The thinner and longer a conductor is, the greater the resistance it offers and the greater the loss of current.

page 26 The load is the bulb, the flashlight body is the conductor and the battery is the energy source.

A battery creates a difference in the number of electrons (and so a negative charge) between one point and another.

Index

atoms 4, 8, 12, 24
aurora borealis 12

batteries 4, 18–19, 20–21, 23, 25, 26, 27, 29
see also car battery, rechargeable batteries,
flashlight battery

car battery 18, 20, 21
cell (battery) 20
chemical reactions 18, 19
circuits 20, 23, 24, 26–27, 29
conductors 24, 25, 26

dam 18

electrical attraction 5, 8, 9, 14
electrical charge 5, 6, 8, 10, 11, 12, 13, 14–15,
16–17, 18, 20, 22
electrical current 19, 22–23, 24, 25, 28
electrical plug 24
electricity
and different materials 24–25
and magnetism 28–29
introduction 4–5
see also static electricity
electrodes 20, 21
electrolyte 20, 21
electromagnet 28
electrons 8–9, 10, 11, 12, 16, 18, 19, 20, 24, 25,
26
electroscope 14, 15

Faraday cage 7
flashlight 26, 29
flashlight battery 18, 20
fluorescent lights 12, 13, 17
fountain 23

generators 4, 26
see also Van der Graaf generator
glowing 6, 12–13
graphite 25

insulators 24

lightning 6, 10, 14, 16, 22
lights, see fluorescent lights, neon lights

neon lights 12, 13
northern lights, see aurora borealis

parallel circuit 27
see also circuits
photocopier 11
plasma ball 5, 17
protons 8–9

rechargeable batteries 20
resistance 25

series circuit 27
see also circuits
static electricity 4, 6–7, 10–11, 14, 15, 22

Van der Graaf generator 15
volts 26

V
79